BIG CATS

LEOPARDS

by Elizabeth Andrews

Cody Koala
An Imprint of Pop!
popbooksonline.com

Hello! My name is Cody Koala

This book is filled with videos, puzzles, games, and more! Scan the QR codes* while you read, or visit the website below to make this book pop.

popbooksonline.com/leopard

*Scanning QR codes requires a web-enabled smart device with a QR code reader app and a camera.

abdobooks.com

Published by Pop!, a division of ABDO, PO Box 398166, Minneapolis, Minnesota 55439. Copyright ©2025 by Abdo Consulting Group, Inc. International copyrights reserved in all countries. No part of this book may be reproduced in any form without written permission from the publisher. Cody Koala™ is a trademark and logo of Pop!.

Printed in the United States of America, North Mankato, Minnesota.
102024
012025

THIS BOOK CONTAINS RECYCLED MATERIALS

Cover Photo: Getty Images
Interior Photos: Getty Images, Shutterstock Images
Editor: Grace Hansen
Series Designer: Neil Klinepier, Candice Keimig

Library of Congress Control Number: 2024938600

Publisher's Cataloging-in-Publication Data
Names: Andrews, Elizabeth, author.
Title: Leopards / by Elizabeth Andrews
Description: Minneapolis, Minnesota : Pop!, 2025 | Series: Big cats | Includes online resources and index
Identifiers: ISBN 9781098246914 (lib. bdg.) | ISBN 9781098247478 (ebook)
Subjects: LCSH: Big cats--Juvenile literature. | Wildcat--Juvenile literature. | Leopard--Juvenile literature. | Black panther--Juvenile literature. | Leopard--Behavior--Juvenile literature.
Classification: DDC 599.755--dc23

Table of Contents

Chapter 1
What Is a Leopard? 4

Chapter 2
Where Do Leopards Live? . . 8

Chapter 3
Leopard Habits 12

Chapter 4
Leopard Cubs 18

Making Connections 22
Glossary 23
Index 24
Online Resources 24

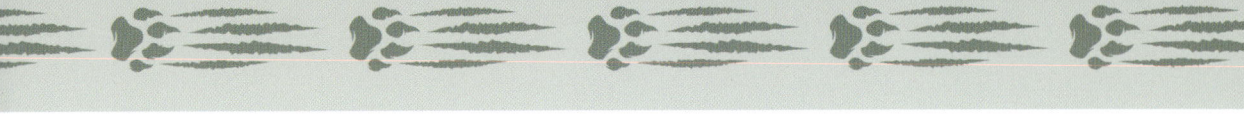

Chapter 1

What Is a Leopard?

The leopard is the smallest of the big cats. Leopards have tan fur with spots called **rosettes**. They look similar to jaguars, but leopards have simpler markings.

Some leopards are all black. They're often called black panthers.

Watch a video here!

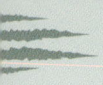

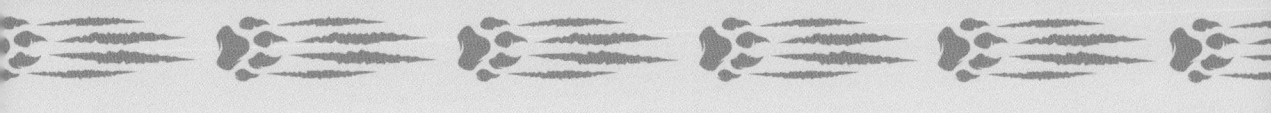

Leopards can be many different sizes. They are usually 110 to 200 pounds (50-90kg). They also have a range of fur colors. Leopards that live in grasslands have lighter fur than those in rainforests.

Chapter 2

Where Do Leopards Live?

Leopards live in Africa and Asia. They live in many **habitats**, such as grasslands, swamps, and mountain ranges. Leopards can live in warm and cold places.

Where Leopards Live

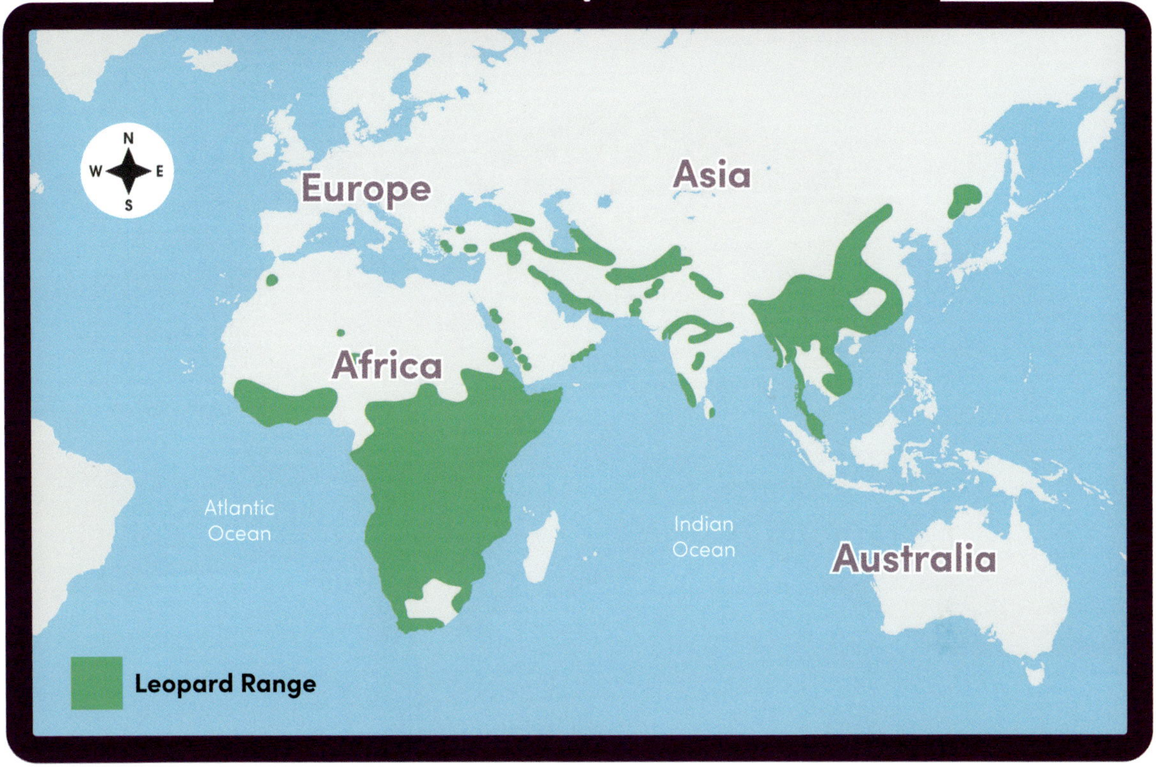
N
W
E
S
Europe
Asia
Africa
Atlantic Ocean
Indian Ocean
Australia
Leopard Range

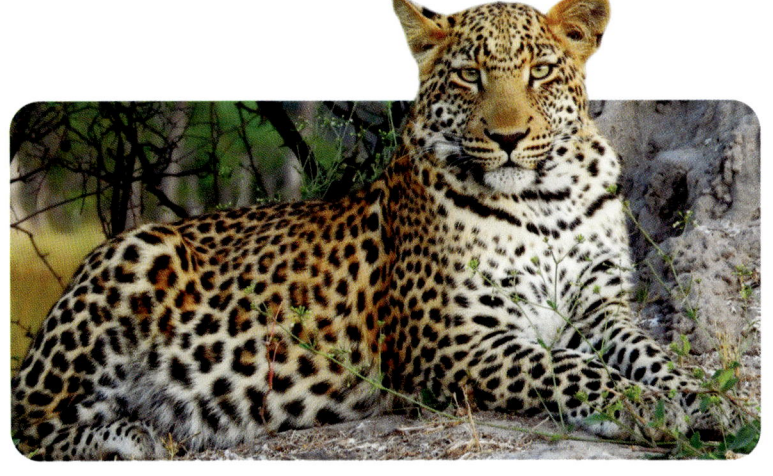

Learn more here!

Leopards can climb up to 50 feet (15.2m)!

Leopards like to live near water. They are good swimmers! Leopards spend a lot of time up in trees. They like to climb. Their long tails help them balance on tree branches. They often eat their food in trees too!

Chapter 3

Leopard Habits

Leopards are **solitary** animals. They mostly live alone. They leave scratches on trees and scent marks to claim their **territory**. They roar to warn other leopards to stay away.

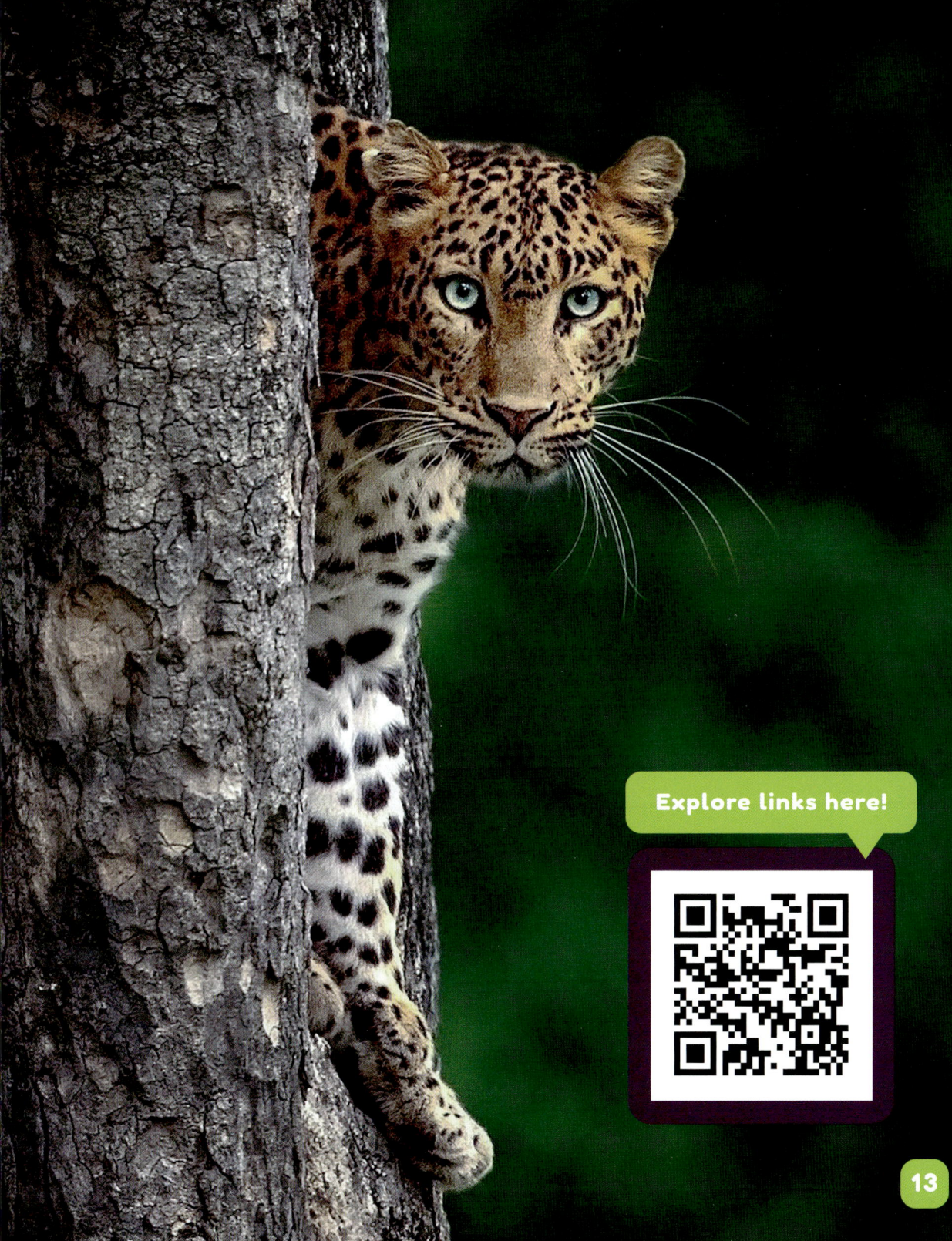
Explore links here!

Leopards are good hunters. They are fast and strong. Leopards can run more than 30 mph (48kph) and leap 10 feet (3m) in the air. They have excellent eyesight and hearing. Leopards rest during the day and hunt at night.

Leopards are **carnivores**. They eat small mammals, fish, and **reptiles**. Sometimes other animals steal a leopard's kill. This is why leopards carry their **prey** high into trees.

> Leopards can go ten days without drinking water! They get most of the water they need from food.

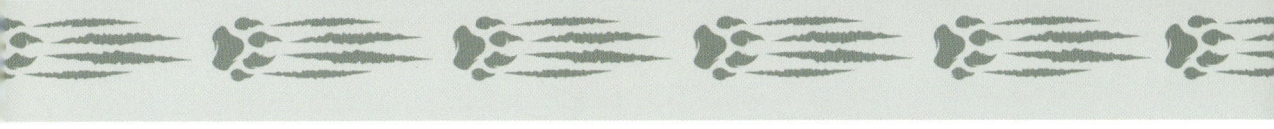

Chapter 4
Leopard Cubs

Mother leopards often have two to three cubs at a time. Cubs are born with grayish fur. Their spots are lighter than an adult's. The spots get darker as they grow.

Complete an activity here!

At three months old, leopard cubs learn how to hunt by watching their mother. They stay with their mother for around 18 months. Then they are ready to live on their own!

Leopards can live 12 to 15 years in the wild.

Making Connections

Text-to-Self

What big cat are you most interested in? Please explain your answer.

Text-to-Text

Have you read about any other kinds of big cats? If so, how were those big cats similar to or different from leopards?

Text-to-World

Leopards look very similar to jaguars. Are there any other animals that look alike?

Glossary

carnivore – an animal that feeds on other animals.

habitat – the home of an animal.

prey – an animal hunted by other animals for food.

reptile – an air-breathing animal that has scales instead of hair or feathers.

rosette – a spot shaped like a rose.

solitary – living alone.

territory – a particular area of land that belongs to an animal.

Index

cubs, 18, 21

food, 11, 16

fur, 4, 7, 18

habitat, 7–8

habits, 12, 15–16, 21

hunting, 15, 21

markings, 4, 18

size, 4, 7

sound, 12

trees, 10–11, 16

water, 10

Online Resources

popbooksonline.com

Thanks for reading this Cody Koala book!

This book is filled with videos, puzzles, games, and more! Scan the QR codes* while you read, or visit the website below to make this book pop.

popbooksonline.com/leopard

*Scanning QR codes requires a web-enabled smart device with a QR code reader app and a camera.